PREPARING THE GROUND

GARY GACH

Heirs Press
657 Mission Street
San Francisco, Calif. 94105

ISBN 0-915970-02-3

Special thanks to the California Art Commission and
The Zellerbach Family Fund for helping make
this book possible.

Illustrations by Bret Rohmer
Cover by Russell Grattan

ENCOUNTER

On the sand I
Got down onto one knee,
Drew her chin closer to me,
Mask to mask,

Yellow dawning on water,
Earthen mirror, or
Fisherman's daughter,
Deity

Whose pores made patterns
Of fish, shells, waves and
When all those lives

Re-focussed her face
I said, "*What?*", & she
Said, "*Who?*"

MARINE

fluid
the dreamfish stride
flat
    like flame
a dream
          vanishes
sideways
           scales
taper to transparence.

surrendering / swimming
swimming / surrendering
waves
currents
spirals
in each
the other
& the big
in the little
eyes
of God floating, flowing.

slow lidless
dance under
water fresh
air comes
in bubbles
rising to moon
light on the waters.

bath
        of forgetfulness
& memory's
                      ancient
crystal
            swirls
back & forth
herds sharing salt
in low tides, tales.

These orphaned children of Pisces
swim in the source of raging
rivers, a sea of Jesus' tears
wept at his immaculate birth
as the first in a long line of children.
Thirst assuaging waters
overflowing Aquarian cups,
these are the fish remaining
after the others I ate up.

may it still come
to what deeper
ancestor

speaking
amidst our narrowing din
& may we be the first to hear.

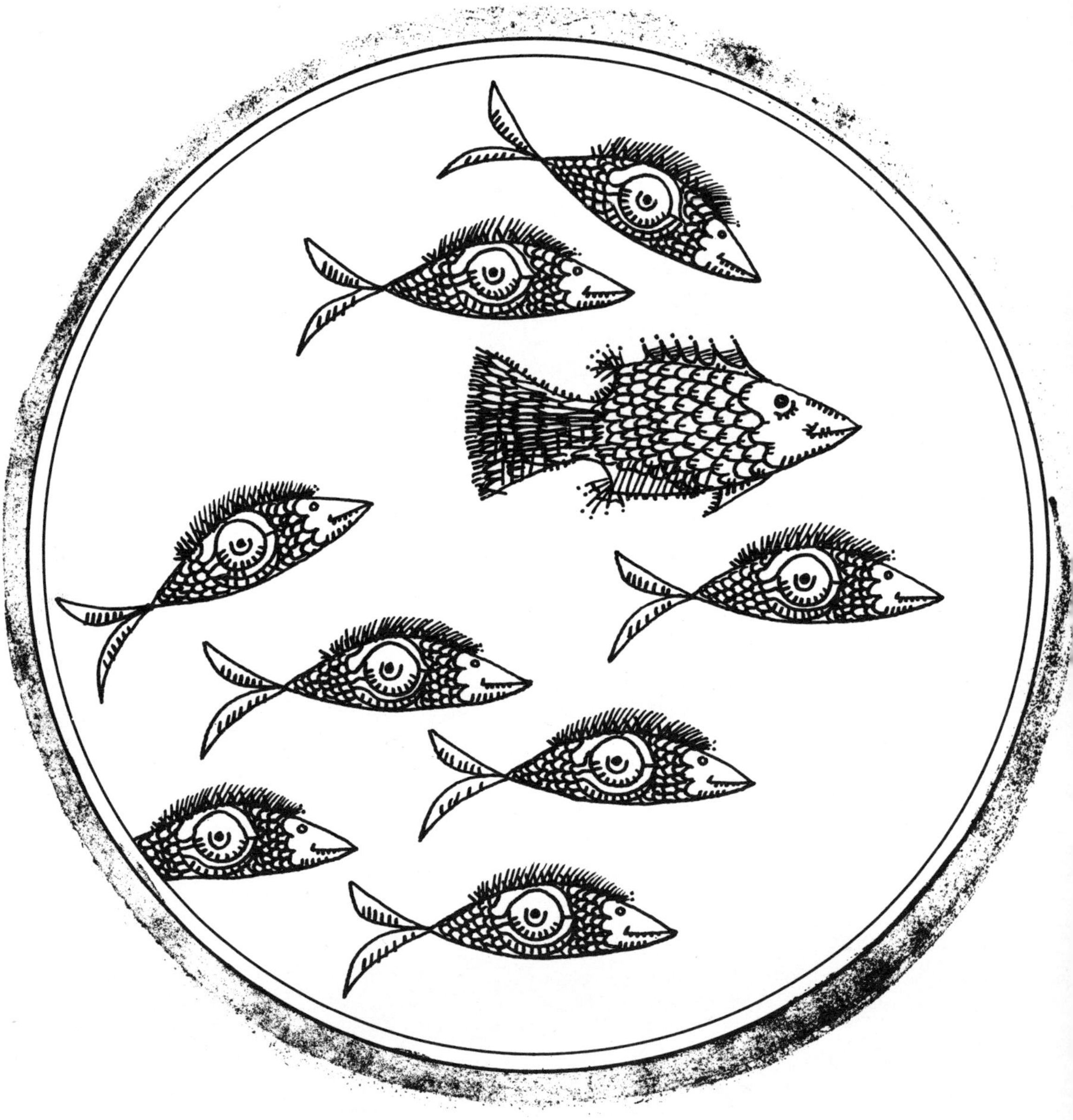

look in some future
mirror & see the fishes
pouring out there as
many as seashores
or our words for them or
dishes.

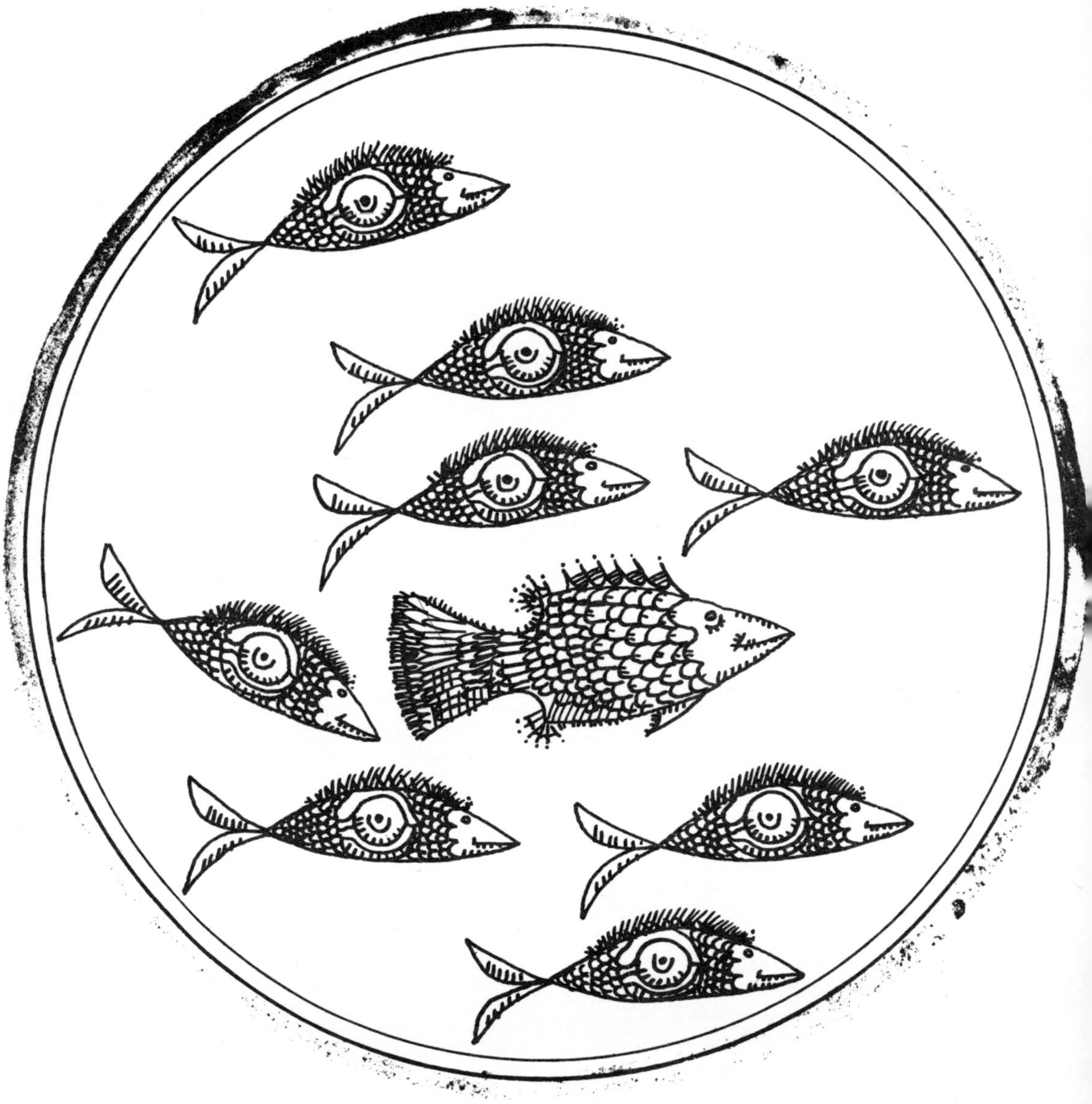

GROUNDING

From what depths risen
Now floating upon   clearing
Voices come to pebbles

In a glade   hand of a statue
Severed from waters we contain
Containing us   diving

Divining, dividing the same earth
Brother, sister, ripple from the warmth of

The dirt underneath the nails of the turtle.

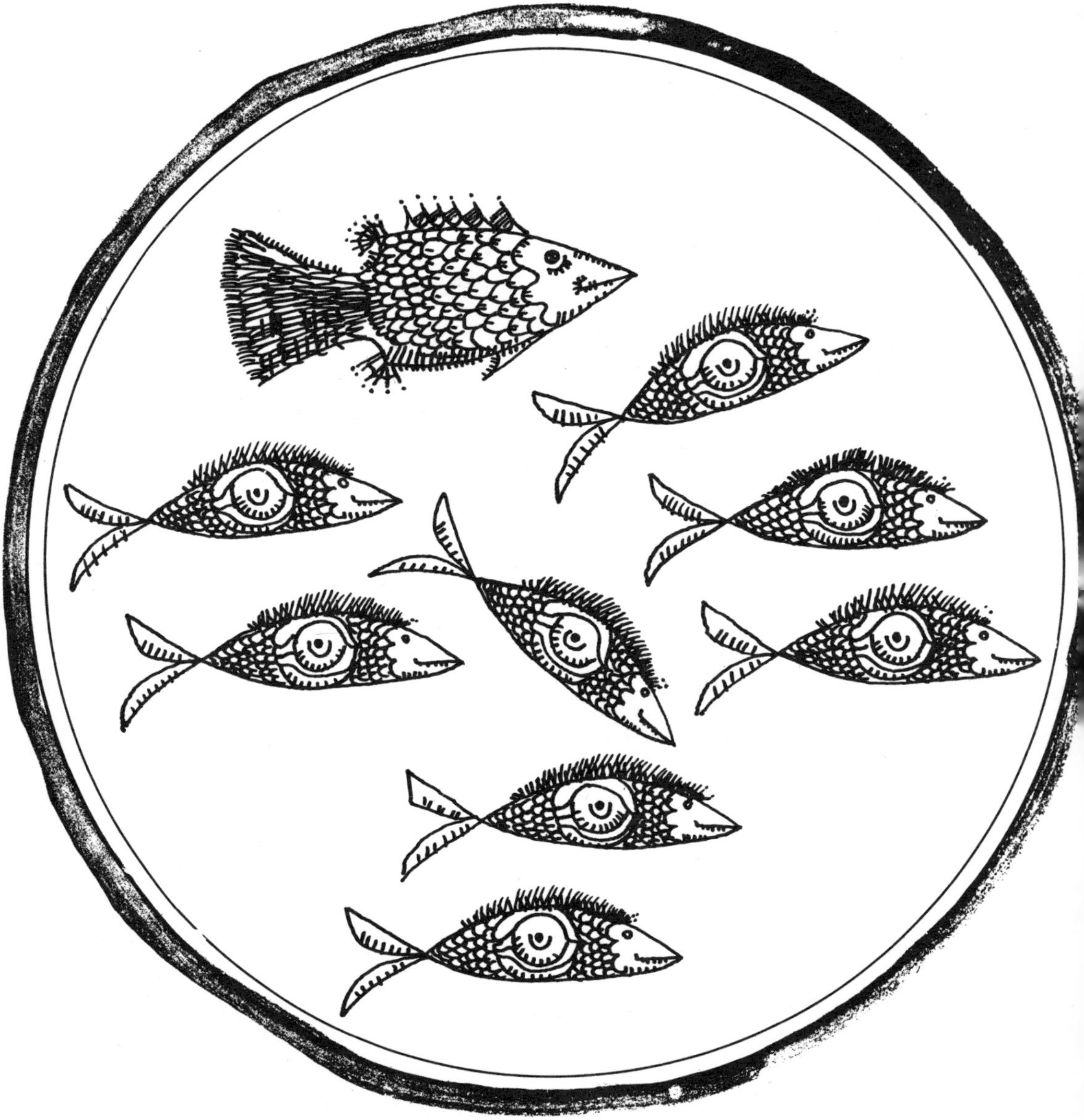

ON AN EPIGRAM OF HERACLITUS

You & I were born playing
Natal tag with dying.
I remember it all.   Before
The streets had actual
Size.   I was it.   You were
It.   All of it.
You ready to go out
Into the world.   I
To come in & begin.

Laid out and straightened
spirals of a shell against my
ear streets lead to and from the bay

petrified beaches of streets

a sound a color a form
of a voice,
water, a bird, a bell

cars over hills like waves

the foghorns, you say, sound erotic

painters outside the building
awaken me whistling bebop tunes

Morning.
People march
To music too unheard and
Vanish just as fast.

The rainbow
You lean on
My shoulder

I encircle at my waist over the
End
Most high most low most beautiful

It gathers
The moisture & motion in me
I reach for

& meet in
My touch in air the words gathering in
Our breaths.

THE MOON

The moon in jagged nets of clouds / looks at you.
A wave splashes against the sand
And alphabet foam looks
At you
The colors of war revolve in air
Looking at you
A pinwheel spins past looking at you
Your mother passes by looking at you
Your sister passes by looking at you
A gull soars overhead looking at you
These lonelinesses would be hard to bear
Were there not loneliness.

MIDNIGHT

he plays harp, stomping
she sings the blues, real
they are one person

listening to them
holding hands I
can get no closer

a radium dial encircles
all the dead ladies
always going home

the unicorn tapestry never
sleeps & the moon rocks on
hold it in   your song.

SPIDER

1.

Our legends tales over looms
Made up & drawn along as
In summer a goose steps

Evaporate dew scorched
Streets of the everyday
The stage. Set. Behind

The wings a lady in waiting
Doodles delicate filament
Angels & devils, cocoons & webs.

2.

The light through trees in fog
A star, perfect, then dispersed in wind

Before eating
You light the candles

Held

Careful not to fall into
Each other's   distraction

Pour the wine
The wind shakes   thinking of

Playing quiet chuckling
Parlor games

The windows   an edge
Alongside those who can't see

The stars winking.

3.

Everything in one instant
Near fatal slip, swaying at mid-point
Of the tightrope & the spectators, the symptom the
Cure, the burst jagged paper disc
Of the hoop dancing at mid-point
In touch almost not there an instant
Across the high lights and greatness with
What is seeking
Return.

LUNAR SEQUENCE

coming from
                    within the earth
                              horizon
space
                    alone
                              still
the need
                    of furling

flags where there is no air,
                         way
beyond
                    the sky —
                              gray dawn,
bright almost as day,
fiery tail smoking
                              the country's
manhood turning on the pike.
                              pink
with lipstick tinkerbelle, kitten,
tangles in the telephone wire.

Inside the wires,
the silver thread,
glimmering

around the earth
           & around the sun
crescent
           & wholly round

precious luminary
           vessel

face
is mirror moon.
hair
           is clouds

stars, mothwings

           & mermaids'
           reciprocal

motion of eyes.

THE MONSTER

The four men talking in the photograph
On the wall of the museum at Culiacán
In formal dinner attire, the camera
Pointing them out

As amidst the rest
Describe a rough circle in their conversation
About us behind their backks
Even after passed.

BEYOND CONTROL

Control lost,
A self, then,

With his drive
To the stars,

Stone to stone,
We know that desire

By now ——

Hitch your wagon
To a star, the

Hitch is, pushed
Close to the earth,

Pulled in, eight
Autumn leaves,

In abandonment, as
She would say they

Are "For Sale",
"Because they give me

Such joy"

NEVER-ENDING NERVE ENDINGS

After skin
Identified with
Childhood situations,

Walls of halls, tubes
And tunnels, turned
On end,

Tower
Long as the path
Of an unshed tear,

Reach for the moon,
And, toppling, form fragments
Of tangents of hands clasped.

neither here nor there —
white clouds drifting in blue sky,
above me below.

thread
Pulling through the buttonhole of my navel
Only so far —

Before I have to remake fabric in its threads
eternity
In my body's image —

Imagination, where
Unsaid promises are, where the beast is
The man in the moon;
Stone, we call

green cheese, an echo
Chamber, cool light, at that tunnel's end,

Whose doors are walls
Our initials carved in conflicting rumors
on bark.

He mocks his loss of
Agony, he couldn't even remember what

She looked like.

The conductor washing his hands in the air
The orchestra mimics the melody he sings.

The world is waiting.

RORSCHACH

1.

In the pool my pores exude I
Dawdle in
                    a drop
Falls and spreads itself out;
Blank page's whiteness;

In snowstorms, over mountains,
Over and over
                            a bear, —
A female bear, a female polar
Bear, — indistinguishable

Nursing her babe at breast all
Full of milk
                        she asks,
"Is it

Good?"

2.

Renewed gaze;
New moon.

Of the visible,
Mirrors and lights.
But of a desire, of
Phases of desire —

The beginning; the half; the
Whole; — old man folds leg on
Knee, leans back, regarding
Full moon

3.

Night as day, heard
Evil spoken of,
Freezing into a body

If not a style

In a world that goes
Around without punctuation,
Just ice, or such, filling in

Parts of an enormous equation

Left blank, at the edge
Of hollow orbits, earthen
Eyes,

Comes around.

and there was a place of water
issuing forth over the stones
and I saw that in one of the places

water pouring over flowing over
the water had carved a heart
in a stone the water carved
  a heart in a stone.

for what we lose
we cannot weep
for what we find
we cannot keep
for always
the star we wish upon
is love.

Between the two head on the third
Above, the wistful arm
Of the fetus, across its brow,

Mast I crane from
To see the sweet foam,
There are other moons,

Other planets, walls, seasons,
Also, always, within,
But one earth

Life of the moon, phantom
Flickering through clouds
Each only once

Smell of fog reminiscent
Of a place, the mind
Slows to a point

Of a thousand views of earth
From a point in space
Each reflecting light

This edition limited to 1000 copies; printed
by Garcia Litho and Printing Service for
Heirs Press in December of 1975. 26 are
signed and numbered by the poet and artists.